Journey to the Source of the Merri

Freya Mathews

Journey to the Source of the Merri

Journey to the Source of the Merri
ISBN 1 74027 197 1
Copyright © text Freya Mathews 2003

First published 2003
Reprinted 2015

GINNINDERRA PRESS
PO Box 3461 Port Adelaide 5015
www.ginninderrapress.com.au

To Maya Ward and Cinnamon Evans

Have you ever had an experience but still, at the end, been unable fully to *imagine* it, to hold it, all in one piece, in your imagination – to *grasp* what has happened to you? This is how I felt after my pilgrimage to the headwaters of the Merri Creek. Memories of the separate days of the journey were fresh, but the poetic significance of the journey as a whole eluded me, even as I burned, bright as a biblical bush, with its after-light.

Where did it begin? Where does any pilgrimage begin? At the dawn of personal time, I suppose, in earliest life, in those one or two images which, as Albert Camus put it, first open childhood's heart. Creeks were already charged with this kind of primal significance for me. A small creek without a name had wound through the back of the farmlet on which I was raised. I had no idea where it came from or where it went, but it was always full of news from upstream, especially after heavy rains, when inflated carcasses of sheep and other animals would float past or pile up against the barbed wires that were here and there strung across its course. These tidings from places that were entirely unknown to me, yet to which I felt mysteriously connected as part of the wider world that the creek created for me, were always tremendously exciting, and I was even more thrilled when the creek would 'break its banks' (how I loved that expression!), and rise up to cover the bottom of our paddocks. What a cold but exhilarating elementalism I felt when I paddled, sometimes up to my knees, in that muddy water that the creek had laid, against the accepted order of things, under fences and gates and across pastures, making the familiar landscape strange.

Of course, it was not long before this elementalism was taken in hand, and the bed of the creek bulldozed out and straightened. Although I was only seven years old at the time, I know I was shocked at the effects of this bulldozing, transforming the creek from easy-going meanderer, so at home in its well-worn, overgrown channel, to raw clay excavation, straight-sided and steep. Still, the billy buttons and watercress grew back, the tadpoles and cranes returned – though there were no more rumours of eels after that – and the creek continued to bring gossip from upstream, and even to break its banks from time to time. My fascinated love for it continued unabated. The creek was not like a paddock or rock or other land form – it *did* things, it *told* things, it was busy and talkative and full of surprises. It was, above all, companionable.[1]

After my family moved from this rural fringe of outer suburbia to Carlton North in inner Melbourne, that rural area was rezoned heavy industrial, and the creek channel was fully 'upgraded' into a concrete drain. No more watercress and billy buttons now. My attention, however, was by this time turning from creeks to other things, and it was not long before I had relocated to London. I remained there for ten years, with nothing but a distant, rarely glimpsed Thames for fluvial company. (I remember, however, my sense of enchantment when I obtained a map of the old Thames river system. Most of the tributaries in the inner London catchment still existed but had been converted into underground sewers. Some of the streets still bore the names of the waterways they had supplanted. Fleet Street, for instance, followed the course of the old River Fleet.)

It was not until I was in my late twenties that I returned to Melbourne, and settled in inner-city Brunswick East, not far from the family home which was still in Carlton North.

That was twenty years ago. It was a couple of years later, when we acquired our first dog, that I really discovered the Merri Creek, which formed the eastern boundary of the municipality. My early memories of this urban creek are blurred. It grew into my consciousness gradually, imperceptibly, as I walked our mad bull terrier along its banks and watched the transformations occurring there. From an utterly degraded and neglected little gutter, winding past the backs of factories and under flyovers, choked with fennel and blackberry, it was returning to life. Under the auspices of its friends (Friends of the Merri Creek) and its heroic committee (the Merri Creek Management Committee), which was coordinating the efforts of local councils and community groups to revegetate the urban reaches, native trees and shrubs were reappearing, and with them the native birds, long since driven from their old grounds, were coming home.

Gradually, a bicycle path was laid, from the mouth of the creek at the Yarra River to the lake at Coburg, and eventually right through to the outer northern suburbs. At the same time, as factories looked for cheaper real estate on the edge of town, community projects moved in on the scraps of degraded 'waste land' the factories left behind. An amazingly colourful landscape started to cohere along the banks. Walkers and riders on the bike path were greeted by the windmills and African grass hut rooftops of CERES (the environment park in Brunswick East), the Aboriginal flag of the Caring Place, and the gold spires and domes of the Russian Orthodox church, which the parishioners were building with their own hands over many years. There was an old market garden, still under cultivation, and re-established swamps and grasslands, and one could pass under the newly cleared arches of bluestone bridges that would look at home in the villages of old Europe.

Ironically, here, where the cause was already lost, where urban-industrialism had already had its way and ripped the innards out of the land, one could find solace and hope for the future. In this small valley, the engines of development that were overtaking the rest of the world were grinding into reverse. While elsewhere, everywhere, cities were expanding, reaching ever further out into the countryside, devouring farms and towns and rural landscapes, the bush was here sending a tentacle back into the city, nature was recuperating its old haunts and, like shoots breaking through concrete, creative forms of new life were springing up.

It was often actually a relief for me to return from the country, where so many pristine beauties were yet to fall, to this place that had already been gutted and smashed, but was now in the process of recovering a new, readjusted wholeness. It had swallowed, or been swallowed by, modernity but had re-emerged, transformed, as a new kind of place, a place in which a new balance had been struck between human and non-human elements.

As I watched this process taking place, the creek gradually became, over the years, my place – my 'country', in something like the Aboriginal sense of that word – and it has sustained me, as it has many other local people, through these times in which our world is being vandalised beneath our eyes.

It was natural then that I should wonder about the Merri's upper reaches. Where the bike path ended, at Mahoneys Road, the watercourse wended off into terrain that was quite unknown to me, most of it privately owned. I had driven up north of Melbourne to look for the creek a couple of times, and gazed at it from rural road crossings, slightly disoriented but also excited at encountering my little friend in such untrammelled guise. But road crossings were few and far between, only four or five of them on the creek's entire seventy-kilometre passage down from the Great Dividing Range to the edge of Melbourne. So its story remained closed to me, a mystery.

When a young friend, Maya, who shared my attachment to the creek, said she would join me if I undertook the journey to its headwaters, the wheels of pilgrimage began to turn. As the idea took shape, it became clear that such an expedition would entail preparations on a scale comparable to those required for more conventional pilgrimages, to India or China, for instance. The project became something more political – an exercise in the politics of repossession and reinhabitation – when we learned that the then state government had put back on the drawing board old plans for a freeway extension along a part of the Merri valley.

Maya and I were at this point joined by a third member of the expedition, Cinnamon, who, like Maya, worked at CERES. Together, these two turned the first stage of the walk into a community event, tying it to the spring Kingfisher Festival (we would walk to meet the first kingfishers flying down the valley to their summer breeding grounds) and to the CERES education program (we would set off on World Habitat Day). They organised a launch and invited hundreds of schoolchildren to join us for short intervals over the first two days. Kind friends and fellow creek devotees agreed to act as our support crew, bringing food and gear out to appointed stopovers at each day's end.

In these and other ways, many people entered into the spirit of our undertaking and helped to bring it to realisation. I came to think that such wider participation is part of the essence of pilgrimage – that the pilgrim carries in her backpack not merely her own longings but those of many others, to lay at the doorstep of the sacred destination.

While Cinnamon and Maya were organising the public side of things, I embarked on a series of reconnoitring trips into the uplands, chatting with local people, setting up stopovers, writing letters to landowners. I had already discovered that another, little known world still existed intact just off the Hume Highway, forty minutes from the city centre, under a patchy overlay of subdivisions and commuter farmlets – a world of bush pubs, pastoral holdings and old homesteads. Somehow, I could not imagine reaching this strangely preserved remnant of colonial days simply by stepping out of my front door and ambling upstream. Could such romance really be within such easy reach? Was it really as much a part of my own world as the hideous industrial imperium that lined the highway? If it was within walking distance of home, then surely it was! But at this stage I could scarcely believe that these worlds could be bridged in this way by my own feet. So I could not really imagine our journey.

There was also something slightly frightening in this as-yet unimaginable prospect. The task we had set ourselves was simply to follow the creek wherever it led – through suburbs and industrial zones and rubbish dumps, under bypasses, beside highways, along the edge of quarries, across bull paddocks and acres of gorse and briar, as well as through any scenes of natural beauty that happened to present themselves. Although the journey was small in scale, and would take us through no wilderness zones nor to any guidebook destinations, it offered the kind of uncertainty that made it feel like a true adventure. For when all frontiers have been tamed and developed, when all exotic tribes and species have been winkled out of their hidden crannies and firmly tagged, where after all can one look for the wild, the unknown? When all natural wonders have been scientifically investigated, and all ancient monuments have become tourist attractions, where can one seek the numinous, the sacred? In a world contracted by motor travel and telecommunications, how can one experience vastness?

Do we have any choice, ultimately, but to turn back to the familiar, to the world we have so rapidly and hungrily made over to our own design? When we examine it again, won't we find cracks in its ordinariness, with strangeness, dangers and distances showing through? Don't we find the wild in the unnamed, the unmapped?

And where are the unnamed, the unmapped, if not everywhere, in everything that has not yet been deemed salient?

Dimensions of salience change. As new ones come into view, old ones recede, tip over the far edge of the past into oblivion. Isn't the unknown forever replenishing itself in this way? Are not wildness and numinosity inexhaustible? 'The Tao that can be named is not the eternal Tao. The name that can be named is not the eternal name.' From this point of view, pilgrimage, as the quest for the unnamed – not to conquer but simply to encounter it, face to face – must be perennial, though reinvented anew by every age. As for distance, what is it but a function of our modes of transmission and travel? To find vastness, haven't we only to set out from our home at the foot of city towers, and in the course of one day's march watch those towers shrink to dwarf dimensions and then disappear? Can one really cover such distances on foot in the space of a single day? This is vastness!

I didn't actually grasp this fully then. But I had had a whiff of the mystery at the heart of the mundane. People seek mystery in the exotic. Every second person one meets these days seems to have toured Outer Mongolia or studied Tibetan scrolls at an anchorage in the Himalayas or been lowered in a cage to meet a white shark face to face in the ocean depths. But when these experiences are commercially preconceived and prearranged, unconnected with the seeker's daily life by a filigree of real relationships, how authentic are they? Can mystery be pre-packaged? Can someone else's blueprint become your adventure? Doesn't the mysterious, this place of transcendent encounter and discovery, lie at the edge or at the source of what is already meaningful for us? Doesn't it lie on the other side of the given? And don't we have to be our own guides in seeking out this other side, since only we can identify the given in our own lives, and find our way to its edges or its source?[2]

So it was that, on the morning of 7 October, World Habitat Day, Maya, Cinnamon and I met at the confluence of the Merri and the Yarra River. Hundreds of children from nearby primary schools joined us as we strolled along the bicycle path in radiant sunshine towards the environment park. Maya gave them the pitch about our walking to the hills to meet the first kingfishers as they arrived from northern Australia.

For seven days we would walk, she said, and the children's eyes widened, as the creek suddenly stretched out beckoningly in their minds.

At CERES a send-off awaited us, with well-wishers and children and speakers from the Creek Committee, the Gould League and Friends of the Merri, all hoping that our pilgrimage would help, in some way, to protect and bless this much-adored creek. Then off we set again, with our jaunty little cavalcade, and no sooner were we out of CERES' gate than a group of children spied the first – and, as it turned out, possibly last – kingfisher of our journey, called up, no doubt, by the kingfisher boogy that had just been sung and danced, amidst squeals and excited kingfisher tweetings, at the CERES cafe.

Through the luminous blossom and foliage of early spring we made our way up past the Caring Place, the golden church, the now scarcely used velodrome, replanted grasslands and swamps, to the old market garden. My friend Joe, who has worked those few acres with his family for fifty years, but now does so on his own with his wooden wheelbarrow and 1957 tractor, was harvesting chicory and spinach as we passed.

Standing in the cool shade of trees that overhang a battered gate, by a large stone trough in which he washes freshly picked bundles of greens, Joe often brings to my mind the old peasant couple at the end of Goethe's *Faust*. In their hut under the linden trees, holding out against the developer, the master of man and nature that Faust in his last years had become, they were the thorn in the flesh of Faustian pride. 'Yon aged couple ought to yield', Faust mutters. 'The lindens still I have to gain,/The clustering trees above the weald/Mock and destroy my wide domain.'[3] Old Baucis and Philemon were mown down to make way for the next two hundred years but, at the end of it all, here is Joe – hemmed in by town houses, yes, but still standing, digging, harvesting in his own field!

How it pacifies my mind to come to this shady spot from time to time, with my basket over my arm, to buy vegetables direct from Joe's weathered hand!

After taking a snap of Joe, we continued along the path, past pigeon lofts, ducks sunbasking in meadows of buttercups, a waterhole where I once saw a giant longneck turtle perched on a post, up to Coburg Lake. Once there, with plenty of time to spare, we camped under a peppercorn overlooking a rocky playground of pools, small cliffs and beaches just upstream from the lake. This was a black duck runway, and watching the ducks plying back and forth across the little amphitheatre, blending in so rightly with the fawns and buffs of the rocks, splashed with red bottlebrush and streaked with blond wattle, the whole scene took on the aspect of a place out of time. Here, just a sliver of a past world of black duck dreamings had somehow slipped out from behind the facade of the present.

'Easy to imagine that this might be/their Melbourne home and increase centre', observed John Anderson, seer of the invisible within the visible, and creek minstrel extraordinaire. 'Nothing more beautiful than the glad charge', he goes on, 'of duck through galleries of red gum – but/here in their absence the cliffs lend that/vista that paces and fledges each dip, makes/the history of that flight visible'.[4]

Oh, that Anderson could have been with us on our journey to see the black duck make their glad charge through the red gums that still actually exist further upstream! But sadly, with so much seeing still ahead of him, he has himself passed into the unseen. Nevertheless, I think of him all the way to the creek's rising.

Ducks and rocks seemed to be the main things on the Merri's mind here, as Anderson would have agreed: it is not for nothing that the creek is called the Merri Merri, as we find out ever more graphically as we proceed.[5]

We cooked a meal that night at the confluence of a storm drain and the main channel, and slept in a scout hall, having been provisioned by my friend Len, the first member of our trusty support team.

The next morning dawned as resplendent as the previous one. We walked springily up the tunnel of willow and ash past Heartbreak House. This is an old lead-lit weatherboard standing in an acre of stone-terraced grounds, with a ruined swimming pool, pines and cypresses, and rambling sheds. I had first discovered it by peeking through a gap in its long, shambling back fence and finding myself inside a veritable Daphne Du Maurier dreamscape. But I had named it Heartbreak House because a friend of mine had come within a whisker of purchasing it, for the purpose of setting up an urban eco-community. At the time that it slipped through our fingers, I had, in my imagination, already moved in, and my mind was full of pictures of myself with my arms around sheep amidst fruit trees and windmills. To walk past the old place still opens up in me little crevasses of loss.

We followed the path past backyard vineyards and vegetable patches, warehouse walls and drive-in screens, to where the valley widened out into a teeming tangle of introduced vegetation. A rather agreeable air of fecund neglect prevailed here, though we did notice that, with the disappearance of native brush, the number and variety of birds dramatically declined. In Reservoir, we were due to meet a group of children from a school named St Joseph the Worker, but we arrived at our rendezvous with a couple of hours to spare.

The map showed a Buddhist temple adjoining the creek nearby, though it was nowise visible from the track. We decided to climb out of the valley to investigate. Away from the creek, the streets were lined with light industrial and automotive workshops. The entire neighbourhood had the bleak, still unlived-in feel of outer suburbs, the buildings punishing to the eye and the truck traffic and wind tunnels battering to the other senses.

Imagine our astonishment then, when out of this harshness a dazzling white colossus, Kwan Yin, Chinese goddess of compassion, reared up, serenely forgiving the desolation surrounding her. She stood, fourteen metres high, complete with neon haloes, in the grounds of an old primary school converted now into the Vietnamese temple shown on our map. Concrete lotuses bloomed in the pond in which she was set, and an open air altar was placed before her. We were stunned by this apparition. I myself was not only stunned but touched, because Kwan Yin, the 'female Buddha', happens to be my own hearth deity, though I had no inkling of her Vietnamese connection.

Beside the altar, there was a monk in saffron robes in conversation with some other Vietnamese folk, and they all smiled as we stood at the front fence, gaping. The smiles broadened when Maya recognised one of the men in the party as Van, a worm farmer from CERES, who was, as it turned out, showing some relatives the Buddhist venues of Melbourne. We were warmly invited in then, and guided around buildings and grounds.

Rough and raw as it was on the outside, the school had been transformed within into a series of colourful shrines and meditation halls, each one devoted to a different ethnic buddha or deity. The main business of deities and devotees alike seemed to be care of the souls of the recently departed, of whom there were numerous small photographs set around altars. Offerings of fresh flowers and fruit were also arrayed in abundance.

Outside, the long scruffy schoolyard led down to the creek, but a little way back from the goddess, set over against the side fence, was another sculpture garden, in the shade of an old white gum. It was named the Deer Park, 'where Buddha first spoke of Darmha to five brothers', according to the sign painted on the adjacent palings; the white figures, larger than life, sat in a peaceful, dappled circle, surrounded by old school benches with sayings, in both English and Vietnamese, such as 'Deluding passions are inexhaustible' and 'Buddha's way is supreme. I vow to attain it', inscribed in gold letters on their backs.

As I was noting down some of the inscriptions, the monk came up for a chat. His name was, incredibly, courtesy of his master in Vietnam, Dao.[6] I told him, as far as language permitted, of my own devotion to the Tao, and he remarked on the closeness of Taoism to Buddhism, especially of course to Zen. We nattered pleasantly about the ineffability of Enlightenment ('the Way that can be named is not the eternal Way'), and also about what it meant to be a monk or nun in a modern context. It meant attaining happiness, he said, not in a temporary or relative fashion, but absolutely, non-contingently. Had he himself attained this, I asked, and he paused, and beamed, and still paused, and then said, not like the buddha, but yes.

He certainly gave an appearance of immense cheerfulness, though he was clearly overburdened with duties, ministering to an entire Vietnamese community, organising ceremonies and large festivals, with only one brother to assist. But I wondered to what extent his good cheer derived precisely from this, this being needed and loved by so many people. Could the happiness be realised without the 'sangha', the community – by someone living lonely and unloved in a rented room, for instance? But given the sangha, would one need the darmha at all – or rather, is darmha ultimately just the training that is needed for maintaining sangha? Whatever the answers to these questions, for a beatific monk named Dao to manifest right at the start of one's Taoist journey, out of the very teeth of industrial disenchantment, was upliftingly auspicious.

After a picnic lunch under an oak sapling next to the Deer Garden, to the accompaniment of pop music from the corrugated-iron depths of the next-door factory, we headed back to the creek. The hundred or so children from St Joseph the Worker duly appeared according to schedule, a little army in yellow tee-shirts and caps advancing across the oval. Cinnamon spoke to them encouragingly about caring for the creek, and then they marched with us for half an hour or so under a hot sun – in many cases on chubby legs alarmingly unaccustomed to walking – along a section of the valley that was still untended, weed-infested. They looked around enthusiastically for native birds, but few were in evidence, and eventually the little troupe departed, flushed but still keen.

We continued along a sealed stretch of path until we came to another patch of bush. In a clearing here we noticed, in long grass, the dome of a large terra rosa rock, somehow reminiscent of those old-fashioned tanks of the early days of deep-sea diving. Closer inspection revealed that this perfect sphere was cast in concrete, and that several others lay, as if they had rolled there, on the other side of the clearing. There were pressings in the surface of these – a few coins and leaves and such like.

The effect was quite entrancing. Spheres, after all, have an ambiguous status: occurring in nature, in cosmological contexts as well as in rocky creeks such as the Merri, they also have the geometricality of artefacts. As natural objects which nevertheless bear this mark of intelligent intent, they can bring an aura of the metaphysical, the mythological, to an otherwise ordinary landscape. These spheres, so unannounced, actually deepened the landscape poetically, rather than, as more often seems to me the case with environmental art, intruding upon it.

After an interlude with a flock of firetails – crimson-daubed finches with lipsticked beaks – who took flight and alighted as with one mind, adding another poetic layer to this haven of bushland still within a day's walk of the city towers, we came to the end of the bicycle path. This is where Mahoneys Road almost converges with the major Western Ring Road. After picking our way through the rubble under the Mahoneys flyover, we emerged into a rather heart-stopping no-man's land between the two elevated, walled highways, with earthworks-in-progress and piles of bluestone making the going rough.

The creek itself of course was undeterred. It nosed around the various obstacles and eased itself under the speedways, though it did become somewhat gagged with rushes at this juncture. But we followed its patient example, and soon found ourselves in the gloom of the ugly tunnel under the Ring Road. The creek ran quiet here, with just some sparse, hardy rushes for company, though Cinnamon heard a lone warbler in those rushes as the darkness deepened. We padded through on a conveniently bulldozed embankment, until we re-emerged into the light of – an industrial horror strip!

This was the moment of truth! We had vowed to follow the creek wherever it led, and now it was leading us into the veritable belly of the beast. Were we up to it? My intrepid companions didn't miss a beat. We pressed on, round the back of warehouses, past dead cars, through truck yards, up into a dirt-track trail-bike maze sculpted out of the hillside with thousands of old tyres, through spiky holes in cyclone-wire fences, past a hill stripped of everything but a broken-down windmill, which gazed out, from earlier rustic days, over vistas of smoking, grinding decimation. It being Sunday, we spotted few if any actual persons, so no one stopped or challenged us and before too long we were able to return to the creek side, which was now waist-high in grass and monster thistles.

And what was this that greeted us? From the very jaws of the city, on the fringes of Campbellfield, the landscape suddenly opened out into a delta of green vastness, a wide valley spreading into a plain that stretched all the way to a green horizon. What joy and relief, like an awakening after nightmare, a vision of what forever lies on the other side of things, the open grassy road that leads back to our beginnings. The thread of the creek wound away across the valley floor, as unfussed as ever, towards an uncluttered skyline.

We beat a path, literally, along the cliff top then down into the wide expanse, leaping a little tributary which Cin christened Pilgrims Creek, until the suburban rooftops of Campbellfield appeared again on the brow of the western wall.

It was now apparent that the valley was laced with dipper-style dirt-tracks, clearly intended for dreaded trail-bikes, but there were few landmarks other than our first river red gum, standing guard over a sandy waterhole, emanating memories of that vanished world in which river reds were the gentle tutelaries of the entire Merri valley. There was also a green pole with spiral mouldings and a ventilated cover that turned in the wind, whispering rustily but eerily. What was it? A religious icon, a prayer wheel? Another unannounced sculpture? A sewerage vent? No need for answers.

We rested, devoutly, under the red gum, watching a black-shouldered kite in the heights above us. I had heard that the good people of Campbellfield had recently launched a black kite festival. Perhaps one day, I dreamt idly, absorbed in the hovering image, there will be linked bird and animal festivals along the entire length of the creek, singing and dancing the world back into its proper semblance.

We stretched out, relaxed. There was peace in the air, but also a certain dissonance. The green pole rustled, the weary old red gum continued its mute witnessing, but the stillness had a sinister edge, as if full of subliminal echoes. Echoes of what? Of the uproar and rage of trail-bikes, bulldozers, chainsaws, axes, guns? Who could know. There was just this uncertainty, these unsettled depths. The only certainty was the creek, always making its way at its own pace – the right pace, as Anderson would say – across the land, attending so faithfully to its effluvial business.

It was now that I began to think in earnest about the creek as Taoist teacher, and creek-walking as a Taoist exercise. Lao Tzu speaks often of the Tao as the soft Way, the way of water: it resists nothing, but in the end nothing can resist it. 'The Softest thing in the world rides right over the Hardest things in the world.'[7] Water seeks out the lowest places, never the high, but by doing so it implacably achieves its goal. Tao's presence in the world, Lao explains, is like a great river or ocean, because, like them, it 'excels at being low'.[8] Elsewhere he puts it like this: 'The highest Excellence is like water. Water, Excellent at being of benefit to the thousands of things, does not contend – it settles in places everyone else avoids.'[9]

A similar point is made in the *I Ching*. Discussing the appropriate approach to situations of danger, the authors appeal to the example of water: 'it flows on and on, and merely fills up all the places through which it flows; it does not shrink from any dangerous spot nor from any plunge, and nothing can make it lose its own essential nature. It remains true to itself under all conditions.'[10]

So the creek finds its way across the land, single-minded but patient, equable, flowing around obstacles, never forcing the issue. It does not strain or stress itself, but follows existing gradients, wandering or simply waiting until a way forward opens, when it quickens effortlessly again towards its destination. Accepting degraded country as unwaveringly as the pristine, the stream parts for car carcasses just as it does for river red gums, and it carries junk as faithfully as it does ducks and ibis.

'Doing nothing' then, in this sense, the creek ensures that everything is done – that the work of the world is accomplished. 'The thousands of things depend on it for life, it rejects nothing... It clothes and feeds the thousands of things, but does not act the ruler.'[11]

Walking the creek, following it along its entire course, really did require, to some small extent, that one enter into this watery spirit. To follow where the creek led – accepting what its course offered, not looking for higher, drier or safer ground, nor taking short cuts to save time – was indeed to acquiesce, patiently and without judgement, in the given. True, we were walking upstream, 'climbing the river' rather than going with its flow, but this did not remove the need for acquiescence, and we could not after all follow the creek, or the Tao, downstream until we had discovered its source, its essence, its true story.

We slept that night in the back streets of Campbellfield, in a local community centre, to which another kind friend, Noel, ferried our food and gear (and gin and tonics – oh pampered pilgrims that we were!). Noel accompanied us next morning back into the valley and a little way upstream to the fabled gorge about which we had all read so much in the newsletters of the committee and the friends of the Merri. Galada Tambore. Here was wildness of a more typical kind:

large boulders, swimming holes and impressive cliffs, sandstone on the east, basalt on the west. Native shrubs resumed their rightful place, and river gum survivors gathered and huddled in inaccessible clefts. A worthy destination for day trippers, if only a path could be cut from Mahoneys Road through all the intervening rubbish!

Afterwards I discovered that it had in fact always been the – visionary – intention of the Creek Committee to bring the path right out to this wide green horizon, thus opening a leafy through-line from city centre to outskirts. This would certainly foster a greater experience of connection with country for the residents of the (traditionally working-class) northern suburbs of our city, deprived as they are of beaches, mountains and major parks. But wouldn't such pathways be desirable in any part of any city, breaking down divisions between city and country, culture and nature, with all the tendencies towards human self-obsession implied in this separation of humanity from the greater continuum?

When, later, I returned home and resumed my usual stream-side rambles, I found that my trip to the rural uplands had brought about for me a transformation of perception. Walking the banks of the creek downstream, but feeling its palpable continuity with the waters upstream, which I now knew, I was actually in psychological contact with the uplands – I no longer felt as trapped in the capsule of the city as I had in the past.

After taking leave of Galada Tambore, we walked all day in a between-worlds zone that was in fact the old freeway reserve, generally a kilometre or two wide, with a skyline of factories always visible at its edge, and an incessant background of machine noise. In the creek gully itself, hawthorn blossom reigned supreme, but there was also an abundance of gorse and, in the paddocks on either side, miles of outsize thistles. The Hume Highway joined us at one point, and we were within hearing of its traffic for the latter part of our way. But how quickly this dark lining to our silver state of mind was forgotten when we came to the native grasslands – one of the last, minuscule remnants of this most endangered of ecosystems in our state.

How absorbed we became in the botanical treasures at our feet. Cinnamon and Maya, far more knowledgeable than I in this connection, put their heads together and identified them. Some I remembered from my childhood, when they were still plentiful on roadside verges, especially the chocolate lilies and varieties of egg-and-bacon. I let the names of others slip by, though I did forge a permanent friendship, I think, with the red-eared kangaroo grass, the original ruler of the Iramoo plain over which the whole of northern Melbourne is laid. (Having encountered the native grasses and wildflowers here in their natural settings, I noticed later, after my return, that I truly recognised them when I saw them in the replantings along the urban banks of the creek – I *knew* them now, in a way I never could have known them through book learning.)

I marvelled too at how scruffy this precious sanctuary looked – how unassuming an aspect it presented to the casual eye, all its treasure in the detail, as is the case with so many Australian landscapes. And, of course, where the native grasses were, there too were the warblers, firetails, larks, raptors, blue wrens and other native folk; at one point I surprised a tiny ground hen sitting on three bronze eggs in a woven nest.

It was late afternoon, and raining again, when we arrived at our destination for that day – Walker's Farm (yes, *Walker's* Farm!), next to a sylvan stretch of creek on Craigieburn Road. Margaret Walker, a free spirit who had found release from suburban conventionality on these seventeen acres, lived here with her parents and Arab mares. The family had invited us to pitch our camp by a billabong at the back of their property. The frogs, roused by the rain, were in full cry when we arrived, and that in itself was sufficient homecoming for me. Jolted by passing traffic, I had not slept the previous two nights, but on this night, while the sky fell in and flamed with one of the wildest storms for years, I slept cradled and smiling in my tent, my ears tuned to nothing but the crooning of the frogs. Our concerned hosts had offered us the stables (yes, oh pilgrims, the *stables*!), but we were safe beside the river red gums, nestled under a stony ridge. We awoke to find the creek in a high state of chatter, bursting with news, and the billabong brimming.

Much later, almost a year after our pilgrimage, I learned that our campsite that night was located in a corner of a sometime sheep station owned by the founder of Melbourne, John Batman. It was reputedly on the banks of the Merri, downstream near the confluence with the Yarra, that Batman had signed a treaty with eight elders of the Wurundjeri-willan clan, the original owners of the Iramoo plain, a treaty that was later disallowed by the government in favour of the legal fiction of *terra nullius*. Batman's flocks, the largest in the fledgling colony, had apparently grazed up here on these wide creek flats in the 1830s.[12] Doubtless the billabong, on the banks of which we had slept so peacefully, was one of the waterholes from which the Wurundjeri had been displaced by the early pastoralists.

From Walker's Farm we set out that morning to follow our ripple-robed old guide past the brick quarries that were scalloping, scooping out, the pinky flanks of Summer Hill. After a wet lunch shared with hundreds of millipedes frantic, for some reason, to get off the ground, we found ourselves finally free of highways and industry, and out into farmlands proper. The valley narrowed here, opening out into little amphitheatres now and then, while the watercourse itself was mostly screened by, and often completely buried under, gorse. The red gums, however, were still in attendance, especially at waterholes. Gracious guardians, spreading their arms out to the water, bending low, limbs entwined, their maternal presence blending somehow with the fabric of the black duck dreamings that seemed to wrap these entranced sites!

As we negotiated fence after fence – scores of them, of every description, generally barbed wire, often electric – we started to work as a team, holding the wires open, heaving our packs from hand to hand, catching each other as we jumped from posts. Our inhibitions began to disappear. There was nowhere we couldn't go, nothing we couldn't tackle! We backed through walls of gorse, hacked our way under railway bridges, forgot about snakes. Dispensing with stepping stones, we simply waded through the knee-high stream when we needed to cross the channel. Anyone looking out from a passing train would now have been unlikely to spot us – we were disappearing into the landscape, squatting under bushes for our tea breaks, acquiring a patina of mud. When we finally arrived at Donnybrook Road, late in the afternoon, we were exhausted but elated. Something was beginning to happen!

We walked the last kilometre to the Mineral Springs Park along a road beside the railway line, while the creek cut through another paddock. We were met, as we approached the Mineral Springs homestead, by a black stallion galloping, riderless, down the dirt road towards us in the twilight. A beautiful white horse was whinnying from a paddock behind us, while a young girl called from the homestead gate. The stallion was visibly torn by the opposing summons but, after thundering along for another moment or two in the direction of the white horse (and us!), he actually stopped in his tracks, turned and cantered back to the black-haired girl, who slipped a rope around his neck, and ran with him, barely holding his bursting strength, back to the paddock from which he had escaped. The girl was Zantha, and there were other equally impressive children at Donnybrook Park. (The horse was actually a gelding, as it turned out, but bred for harness, which explained his imposing appearance.)

The park had recently been refurbished by the children's father, Jerry, who was trying to revive this old pleasure-ground, site of the first piped mineral waters in the state. There was a lot of magic here, flawed, slightly out of date, but with its peacocks, parrots and geese, it was certainly reaching for the indefinable, for the spirit of the 'lost domain'. Its effect on Zantha and her siblings, Chelsea and Gabe, was striking: with their innocent but straight-to-the-heart-of-the-matter questions, they had stepped from the pages of some enchanted tale, but they seemed in consequence more fully present in the world than most of their media-mesmerised contemporaries. We were warmly welcomed, served tea and pavlova from the previous day's wedding reception, and offered refuge from the night's rain on the floor of a lumber room – an offer we gratefully accepted.

Pampered again by another member of our support crew, Cathie, we listened while awaiting dinner to Jerry's stories of the district. Nearby Kalkallo had once, it seemed, been a watering place for bullock teams. Originally named Rocky Water Holes, it had served literally to water the bullocks but figuratively their drivers too, with six or seven pubs in the tiny settlement. The drivers would hole up there on the flood plain of the creek and wait until a sailing ship hove to in the port of Melbourne town. An old sea captain with a homestead on Mt

Ridley, north of Craigieburn, could see all the way down to Port Phillip Bay with his telescope, and would let the drivers know when sailing ships appeared.

I had sensed a bit of this frontier pub tradition when I had first pulled in to Donnybrook on one of my upland scouts, several weeks before. After the drive from town along the Hume Highway – which seemed to hold, in its forty hard-hitting minutes, all the crass aggressiveness of late modernity – I had been astonished to wander into the Donnybrook pub, there to discover, in the persons of Graham and Morrie, the veritably pre-modern ambience of outback Melbourne. Chatting with these lifelong local boys, I could have been in a pub in the back blocks of the Pilbara. They told me about old mills and Chinese market gardens, dams and swimming holes, along the creek. I was struck, as I would be more and more forcefully in the course of our journey, by the historical richness of this route.

Later, after our return, I discovered that the explorers Hume and Hovell, the first white men to cross the Great Dividing Range, had followed the Merri valley for a day or two on their way down to Port Phillip Bay! How strange to think that our very own creek had been a conduit of colonisation in this way. What to make of this equivocal discovery? I was pleased to think that though we were partly retracing this march of colonisation, *we were doing so in the opposite direction!*

As we left the park the following morning, we stopped for a ceremonial cup of the spring waters, trickling from a rusty pipe into a stone recess. They tasted metallic but sweet. Delicious! The special distillation of our creek. Holy waters indeed! I vowed to return to collect it in bottles. If Joe's vegetables, grown on the Merri flood plain downstream, were my soul food, then this would be refreshment for my spirit!

Now we entered a seemingly charmed zone, a park-like reserve and pinery, with lawn down to the water's edge. The creek was at its most alluring here, with stone terraces and amphitheatres, steps and water gardens, shoals of shards and spheres tossed together, like simple petrified souls. How busy, how talkative, the creek was amongst all these rocks! 'The Merri Creek saying the right thing/over and over', as John Anderson had noted.[13] Meadows of purple iris bloomed between the river gums. The pine trees, however, lent an edge to the otherwise cheerful atmosphere, as if something were present in the shadows, watching, with an indefinable attitude. A wedge-tailed eagle crossed our path, very close, and the vague sense of menace was underlined when a falcon dived past with a tiny bird – perhaps a blue wren? – in its talons.

We were now approaching the foot of the long-visible Bald Hill, from which, my later reading revealed, Hume and Hovell had first viewed this entire southern portion of the continent. In his diary, Hovell had named it Perambulator Hill, because the pram device he was laboriously wheeling with them to measure distances had here broken into pieces. Hovell wrote of the sight that met their gaze from the hilltop: 'this was plains, and open forest, which served to give them a more beautiful appearance, beyond the reach of the Eye and as far as we could see With the Spy Glass (say) from S.E. to West and the land falling with a graduel decent towards the South, never did I behold a more charming and gratifying sight, at least not were it in its Natural State.'[14] Others were to share Hovell's enthusiasm. This was the site of the future city of Melbourne.

At the foot of Bald Hill, the railway reappeared, bisecting the landscape into separate countries. On the west side of the weathered bridge – the side from which we had come – the land told a gothic tale, full of beauty and suspense, a product of the potent interweaving of European and Australian elements, faerie overlaid on dreaming; on the east side lay a wide plain of native poa grass, innocent of secrets, open to the sky, the creek no more than a line of river gums scrawled towards the horizon. The channel of the creek was full, flush with the grass, long ribbonweed tresses loosed in its stream. Even the rocks had vanished, though the sheep remained. A little way along, scatterings of lace-trimmed rocks reappeared, and a particularly celtic-looking mound, studded with worn old rock teeth, beckoned to us.

Perched on the crown of Pilgrims Knoll, we were treated to an amazing show of speeded-up weather: at first storminess impended, and Bald Hill blackened; then it was as if the sky relented, breaking out into forget-me-not smiles and gilding the hillsides, only to furrow over, dark and beetle-browed again. Enthralled, we sat and applauded these atmospheric theatrics.

By and by we recrossed the creek, avoiding bull paddocks, and continued for some time through an amiable emptiness. Our feet, however, were by this stage rather sore. Maya had resorted to placing hers in plastic bags inside her boots, in an effort to keep them dry. My toes were the shade of ripe plums, and one of them was turning black. With shopping bags bunching out the tops of socks, trousers rolled up, collapsed brims of sodden hats, and a repertoire of funny walks that favoured our fitter parts over our failing ones, we must have looked a sorry little band of water rats.

The barbed wire fences loomed relentlessly. We swung packs, scaled posts, squeezed through spiky gaps, did the limbo at a dozen different heights. Finally we arrived at a track that led to what we hoped was the farm at which we had arranged to spend the night. We broached the flooded ford, and Cinnamon rescued me when, mesmerised by the knee-deep, fast-moving water and top-heavy with my gear, I kept gently capsizing, surrendering to the flow.

So it was that, decidedly wayworn, we trudged up to the cluster of roofs and hedges at the centre of an immaculate estate, trying, as we went, to stay out of the enclosures of all manner of over-frisky-looking livestock. We did eventually, after various false starts, find the 'cottage' (a tastefully furnished, five-bedroom farmhouse!) in which we had been invited to stay, and we sank gratefully into the floral sofas. We decided, then and there, that we needed a rest day, and when Kate, our generous host, appeared, she agreed without a blink to our request. A whole troupe of our support team soon arrived, with hot food that would have moved less susceptible hearts than ours to sighs of gratitude, and the contemplative pace of the day changed abruptly to one of chatty conviviality.

We bedded down contentedly that night under fresh doonas in separate rooms, and I fell asleep wondering at the ancient tradition of hospitality towards the stranger. Even in cultures such as ours, in which this tradition has so thoroughly lapsed, it resurfaces spontaneously, it seemed, as soon as an opportunity arises. In most cases we had only had to ask, and people had offered us their paddocks, stables, spare rooms, their showers and, in this case, their vacant cottage. Our hosts had been people of widely different backgrounds, yet each had responded, in their own way, without hesitation and with utmost graciousness, to the archaic pilgrim call, a call so counter to the demands of possessive individualism that frame modern life.

The following day, while nursing my purple toes, I had leisure to ponder further the meaning of our journey. That it was indeed a pilgrimage was clear, and the meaning of pilgrimage itself – in contrast with bushwalking, for instance – was now coming into sharper focus. For bushwalking, in its more dedicated forms, is surely a pursuit of unspoilt nature, an exercise in withdrawal from human relativity and mediation, the better or more directly to experience the ground and un-negotiable context of human existence. Such a withdrawal from society seems necessarily to require an ethos of self-sufficiency and physical austerity.

Pilgrimage, by contrast, was shaping up as an exercise in the acceptance of the given. After a single initial choice – that of one's sacred destination – one hits the road. Once on the road, however, one takes whatever comes. There is no preference for the non-human over the human here, nor is there any sense that the human is less a part than the non-human of the numinous, the sacred, the absolute order. One rejoices in nature and the open skies, but trudges through industrial darkness and urban labyrinths if that is where the way leads. One's dependence upon others is acknowledged, and the alms of rich and poor alike are gratefully accepted. Tonight might be spent crouched in a wet paddock, but tomorrow night one might be plied with gin and tonics, sitting up, conversing brightly, in front of a blazing lounge room fire. In this respect, pilgrimage seems, again, to be Taoist in its essence: one follows the Way trustingly, resigning judgementalism and control, accepting succour from any quarter.

And what does the pilgrim offer in return for this sacred hospitality, I wondered. Well, she *gathers in*. Her journey gathers the people and places she encounters along the way into a sacred story. By journeying to her destination – traditionally a temple, shrine or other holy site – she draws the entire landscape and its inhabitants, including her benefactors, into the net of its meaning: through her intention, her dedication to that meaning, she binds everything into her wake. In our case, we were gathering all that we encountered into the narrative net of the Source of the Merri, threading people and places, animals and birds and plants, landforms and weather to a single spool, winding them into a poetic unity.

However, though it was true that our journey was undertaken in a pilgrim spirit, it did not coincide exactly with pilgrimage in its traditional sense. For while any devotional journey to a sacred destination may be described as a pilgrimage, the destination of the traditional pilgrim typically lay in foreign lands. (This is reflected in the etymology of the word 'pilgrim', derived from the Latin *peregrinus*, foreigner.[15]) The sacralising force of pilgrimage, moreover, is not tied in any intrinsic way to landscape itself: a philosopher, for instance, might undertake a pilgrimage to Athens to honour Socrates. His journey would then indeed gather all the elements he encountered along the way into a narrative, but this narrative would knit these elements into a story about the sacredness of philosophy rather than the sacredness of the landscapes he had traversed.[16] Our walk was a pilgrimage, but it was not a journey 'to foreign parts'. It was on the contrary a journey into our own country; the land itself was our destination and our object was to gather *it* into a poetic unity and thus to make it more our own. Moreover, while a wide variety of modes of travel may serve the purposes of pilgrimage, it seemed that *walking* was integral to the spirit of our endeavour. Our step-by-step contact with the land created an intimacy that bound us to it and it to us.[17]

As I mused on this, sitting in a cane chair looking out over the well-tended pastures of Merri Park, a whole tangle of different kinds of journey started to distinguish themselves in my mind, and our own walk seemed to partake, in some way, of each of them. A pilgrimage it certainly was, but it also seemed to tap into the spirit of ancestral journeys – the creative journeys of the mythical ancestors, or 'dreamings' of Aboriginal peoples. These ancestors travelled across empty, primal landscapes, differentiating by their activities dunes and salt pans, rocks, rivers and waterholes, eventually creating further landforms and totemic species by their own metamorphoses. Such dreaming stories – stories of long-ago mythical journeys – gather the present landscape into poetic unities for the peoples who inhabit them.

Dreaming stories are, however, in a continual state of evolution, with new stories emerging, so that the poetic unity of the land is perpetually in process of renewal.[18] The unifying function of the stories is reaffirmed by the fact the journeys they describe are re-enacted by the people to whom they belong. Aboriginal people follow the steps of the ancestors, singing and dancing the stories of the primal journeys, reinvesting the land with meaning and coherence, bringing it to life. When the world is thus sung and danced to life, it is possible for human beings to enter into communicative relation with it. Through such journeys, then, we can truly inhabit our world, belong to it, become its natives, its people.

When journeying is understood in this way as 'gathering' everything in the journeyer's wake into a poetic unity, waterways themselves take on the aspect of prototypal journeys, or journey-lines – the journey of water through land to sea. Each creek and river binds the elements of the landscape through which it passes into the story of its own adventure. The landscape is sculpted to this adventure, all its features touched by it and subtly turned to it.

Like pre-existing songlines, then, the waterways knit the land up into narrative unities. It is via such unities that distinct 'country' comes into being. But while waterways allow for the discrimination of one 'country' from another, they also *link* the 'countries' they serve to individuate. These watery storylines thus organise country via a simultaneous action of differentiation and connection: they draw the world into wholeness while at the same time generating difference.[19]

In walking a waterway, then, we are already repossessing all that it has gathered into its narrative terrain. In retracing its primal story, we not only renew that story, but expand it, in our turn gathering all that we encounter into the story-net of our own journey.[20]

In thinking about ancestral journeys, however, I also had to consider how our expedition compared with those other journeys with which our continent is so liberally inscribed – the epic journeys of colonial explorers, such as our precursors, Hume and Hovell. On the face of it, journeys of exploration also served as 'ancestral journeys', but they did so for the colonising cultures as opposed to the indigenous ones, marking with dramatic narrative features maps which were as yet, from European perspectives, almost as blank and undifferentiated as were the primal dreaming landscapes of Aboriginal provenance. These journeys of exploration enabled the colonisers to settle the land at the level of imagination as well as at a material level. Moreover, by risking and sometimes losing their lives, the explorers symbolically staked out a *moral* claim to the terrain their journeys 'opened up': by their real or perceived heroism, and the epic scale of their hardships, they 'won' this new terrain, mythically, for the conquerors.

However, the relation to land that was established by these journeys was of course one of conquest rather than of mutual belonging. The explorers were not pilgrims. Their journeys did indeed gather the people and places, the species and landforms they encountered, into stories, but the reference point of these stories – the centre around which they turned – was not a sacred destination but the personality of the explorer, the 'hero', himself. If pilgrimage draws everything the pilgrim encounters, including himself, into the net of meaning provided by the destination, the colonising journey draws everything the explorer encounters to himself – he arrogates the landscape to himself by the force of his heroic will.

Instead of walking the world into being, then, bringing it all to life in the wrap of meaning that emanates from a sacred destination, the explorer walks himself into being. He enchants and charges up his own personality (and by implication, that of his race and class and culture) by *overcoming* the people and places he encounters, mythically absorbing their life force into himself, rendering them objects of his own triumphalism.

But what distinguishes a journey of reinhabitation from one of colonisation? When the children of the colonisers want to belong to the land, rather than have it belong to them in the manner of their fathers, is this merely a further extension of the process of expropriation? If the new, would-be natives ignore the history of the land they wish to reinhabit, then perhaps it is. But if they shut their eyes and ears to the story of the land in this way, then they cannot fully reinhabit it in any case. To belong to the land is to uncover its layers, discover its story and weave one's own identity into that story. In our case, here in Australia, this story includes chapters of bloody invasion. We cannot enter the land without bringing these chapters to light and acknowledging them. To acknowledge them is also to deal with their aftermath, to seek some kind of justice or redress for those who have survived.[21]

As an exercise in repossession, it also seemed important that our journey was largely through private land. For the subjugation of land by modern societies, and the consequent alienation of culture from 'country', is expressed most forcefully through the institution of private property. Private ownership, as it is currently understood, nullifies the mutuality between 'landholder' and land: the land belongs to the holder, but the holder does not belong to the land. Even more importantly, while giving legal control of a portion of land to one or several parties, the institution of private ownership generally excludes the wider public from accessing it. 'Trespassers Will Be Prosecuted.' Under modern property regimes, the world is no longer in any moral sense *our* world: access to most of the places in our immediate neighbourhood is drastically curtailed. How many of the houses in my own street, let alone in the whole of Brunswick, have I seen inside? How much of the Merri Creek in its upper catchment is even visible, let alone accessible, to the wider public?

This physical exclusion of people from their homelands began in Europe long before it was re-enacted on the indigenous peoples of the European colonies. We were of course all 'indigenous people' once, and not so long ago, but we Europeans were dispossessed by way of the privatisation of the commons – the centuries-long process known as the Enclosures – prior to, and concurrently with, the dispossession of indigenous peoples in colonised lands. Many who were dispossessed on their own soil became dispossessors on conquered soil, but the process of dispossession, whether in old lands or new, has always been a corollary of capitalist economics, and continues unabated into the present.

This process is perhaps less visible in a contemporary frame, because people generally are today *born* into a condition of dispossession, in the sense that they are from birth denied physical and spiritual access to the greater part of their immediate environment. This deprivation is eased for some by their own exclusive possession of a house or apartment, but many lack secure access to *any* portion of their surrounds, and most, even those who do enjoy some degree of private ownership, are profoundly impoverished in their relation to the larger world. 'Home' refers, at best, to a small plot, and for many this word has no tangible referent at all.

Compare this state of metaphysical dispossession with the wealth of indigenous peoples of pre-modern societies and eras. I remember once standing on the tip of Cape Leveque in remote north-western Australia, looking out across a string of pink islands of unearthly beauty in a live, primeval sea. This was the 'backyard' of the Bardi people, with whom I was at that time staying, a small group of families who, for thousands of years, had hopped from island to island on ceremonial business, swimming the shallow channels, escorted by tiger sharks through waters plied by dugong and sea turtle. This wonderland had been *theirs*, their home, their world, the very substance of their soul. And what was *mine*? A tenth of an acre in Brunswick? What did this say about the dimensions of my soul? And I was one of the lucky ones!

Even medieval serfs were in this respect better off than most moderns – the world through which they walked was one of wildwoods and commons, spiritually open to them, even if sometimes physically out of bounds on account of bandits and wild animals. Or take the camel nomads of northern India. Until very recently they had the free run of their ancestral migration routes, even though these lands were legally owned by others. Traditional property regimes in this part of the world were regimes of usufruct, allowing for multipurpose, communal access to land, with no one party having exclusive moral title to it. Different groups and peoples were able to participate fully, on spiritual and imaginative levels, in their immediate world.[22]

To be at home in the world, to experience one's immediate environment as one's *own* world – isn't this a basic metaphysical right of all human beings? Isn't denying people moral or spiritual ownership of their world cutting them off at the knees, psychospiritually speaking, condemning them to a most abject metaphysical poverty? Yet this metaphysical poverty seems to be a routine consequence of modernity. Doesn't it show up in a poverty of poetic expression at the level of everyday life – a poverty of expression all too apparent in modern societies?

Compare daily life in industrialised modern cities with the daily life of those north Indian nomads, for instance. The almost exclusively utilitarian tenor of the modern city, with its resultant ugliness and discordance, is obvious to anyone standing on a busy Melbourne street, whereas an equally casual glance at a nomad camp reveals a form of life expressive, down to the last detail, of a poetic sense of existence: the most mundane activities are 'raised', through song and story, costume and dance, to a form of poetic address, an ongoing call to a responsive, mythically activated world. We moderns may be offered abstract or transcendental compensations for our metaphysical disenfranchisement: we are invited to ground our identity and our imaginations in philosophical, religious, nationalistic or other ideals. But aren't these pale substitutes for the inexhaustible reality of a living, speaking land, a land that greets us as kin as it stretches away and away in every direction, filling our senses and emotions with a richness of experience that no amount of clever abstract thought could ever provide?

To urge the repossession of privately owned land through walking is not to waive the legitimate rights of landowners. The pilgrim will surely seek the owner's permission to cross their land, and will treat their pastures, fences, livestock and so on with all due respect. She is legally a potential landowner herself, and landowners are in any case, in a larger sense, like everyone else in the cultures of modernity, relatively metaphysically dispossessed. There is thus no us/them divide in this equation. By harbouring the pilgrim, the landowner is acknowledging an older and truer compact between human peoples and the land, and thereby helping to ameliorate the metaphysical pauperisation of our collective life.[23]

Of course, it was not possible to ask for permission to enter *all* the properties that adjoined or were crossed by our creek, and it has to be admitted that the fact that this land was conventionally off limits gave our journey a special frisson: we were venturing into places we would never previously have entered, and passing points at which we would once unthinkingly have stopped. There is danger in giving up the safety of remaining in just those spaces one is by social convention permitted to occupy. If any harm befell us as a result of such trespass, we would receive little public sympathy. And privately owned lands are in a sense a last great frontier, for there is no public record or account of what goes on, or is kept, within such grounds. This lent that edge of unpredictability to our progress that turned our walk across tame farmlands into a trek into a true wild.

It was clear that, in the course of this short journey, we were beginning to lift the lid on a rich brew of meanings and traditions largely ignored by modern societies. But it was not only the meaning of journeying that we were rediscovering.

Back at Merri Park, the term 'day of rest' was taking on fresh meaning as well. It wasn't quite a sabbath, because it was our sixth rather than our seventh day, but we were certainly in need of re-creation. I basked in intervals of warm sun in my cane chair while Maya and Cinnamon went off on botanical business. Magpies mewed in the grass, cockatoos shrieked as they passed, unseen birds chattered to themselves in piping voices all around the house. Our boots and gear were spread over the yard to dry.

Later, Kate came down from the main house and told us a little about the property. The entrance drive was lined with oaks grown from acorns gathered in the grounds of Buckingham Palace. This formal avenue set the tone for the front of the estate, but over the ridge, out of view of the road, the land reverted to its natural rocks and river reds as it descended to the creek flats. There was a lot of love in the air here. A commitment to land care, to the return of native grasses, to replacement of the dowager red gums. Subdivision was, apparently, inevitable. Farming is uneconomic and, when town water comes to these parts, large landowners will be 'rated out'. My heart sank at the prospect of this country being subjected to yet another wave of appropriation, but it was some comfort to remember that hobby farmers can in fact prove to be sterling conservationists, establishing more habitat for native species than broadacre farmers possibly can. We considered what it would take to establish a reserve along the entire length of the creek: the consent of all landowners, and $10,000 per kilometre for fencing. Something in the order of $700,000 in all. An impossible dream?[24]

Revitalised, we set out early on the drizzly morning of the seventh day – without glimpsing the legendary white-tailed water rat that lives under the Merri Park bridge. The creek meandered through Scottie's bull paddock (a matter of some concern for me!), then it cuddled up to the road for a way, before striking out across quite empty paddocks, past the back of Barrows' farm. The rocks and river reds had almost disappeared here, and the creek ran bare through grass till it led right up to the back door of a dilapidated farmlet. An assortment of black-faced animals – goats and geese and sheep – greeted us from tumbledown sheds, but no one else appeared. The fences, improvised out of old door frames and bits of tin, proved almost impassable, and it was only after some fancy acrobatics that we emerged onto the Beveridge Road. There we crawled under a wet bush and sat peeping out into the rain like Ratty and Moley in *Wind in the Willows*, sipping our tea in surprising snugness.

The landscape was bleak at this point, but it was also profoundly charged with the proximity of Ned Kelly's house. I had been astounded to discover the existence of this house on one of my reconnoitring trips. Kelly had, it turned out, been raised on the edge of the hamlet of Beveridge, a couple of kilometres west of the creek. I had driven over to the town – presently, inevitably, in the throes of development – and followed signs to the house. Imagine my delight when I found, not a touristified bit of historical kitsch, but a genuine ruin, an early colonial weatherboard with characteristic low ceilings and hand-hewn timbers, its roof collapsed, its boards dangling, patently unsafe to enter. Oh joy! Only the bluestone chimney had been restored. Rotting old hawthorn bushes still blossomed wildly in the yard. A sign on the cyclone wire surrounding the site said that the house had been built by John 'Red' Kelly, Ned's dad, and that the family had occupied it until 1864.

I stood at the back of the house, and pictured little Neddy playing there. The country to the south and east – the direction of the creek – was still free of housing development, and I tried to imagine it through Neddy's eyes. He would have known the Merri well, given the affinity between boys and creeks, and he probably fished and yabbied there.

Back at the side of the Beveridge Road, after our tea break under the wet bush, we found ourselves in a huge paddock, with just a smudge of gorse marking the line of the creek. Soon the ground started to rise, and the rocks reappeared, and with them the hawthorn, still in exuberant bloom. This made presently for a very pretty landscape, in a moody, moorish sort of way: rocky mounds laced with lichen and bowers of creamy-jade hawthorn marked the watercourse as it wound across wide pastures. Sheep completed this celtic tableau, though ebony cattle also appeared, and I finally found myself face to face with a bull – though a youngster, fortunately. He stared big-eyed from his blossomy bower amongst the boulders, and followed me with his look of fixed astonishment, but did not budge.

Like an emanation of this celtic scene, a homestead now appeared on a ridge up ahead, secluded amongst pines, but rivetingly picturesque. On the map it was named 'Camoola', but 'Wuthering Heights' would have been more apt. The creek led us right through its cluster of antique buildings, which included an early bluestone farmhouse, windmill and huge, two-storey, bluestone barn, small windows set in its high walls. A vehicle stood in the drive, but no Heathcliff appeared, so we trudged on, through gates and yards, past more ebony cows, and out onto another, even more expansive plain.

The only landmark on the entire horizon here was a rather odd kind of pointy hut, set like a sentry box in the midst of emptiness. As it was still raining, we made for this apparent shelter, as a likely lunch spot, but on closer inspection we found it to be a rather greasy machine shed, ringed, in good fairytale style, by a moat of mud and a circlet of ferocious thistles. Maya, ever game, improvised a drawbridge with a couple of planks but, when she stepped on them, they sank. Stranded at the entrance to the grimy tower, with no prince-to-the-rescue in sight, she had no alternative but to splash back glumly through the mud, adding sludge to the cold slosh that already filled her boots.

After more trudging, we spied (yes, we were definitely by now in fairytale mode) another cluster of intriguing gables in the distance. This also turned out to be a 'lost domain', tucked amongst trees and hedges, with rose-coloured lofts and roofs. On this occasion, however, the creek did not lead to its front door, so we contented ourselves with gazing at it from a grove of gums in which we spread our lunch. As it turned out, however, this was no ordinary lost domain: I was stunned to discover a few weeks after our return that 'Walnarring' was reputed to be the actual birthplace of Ned Kelly, the site of the original Quinn homestead. The Quinns were Ned's mother's family, and the whole Kelly drama had originated here, on the very banks of the Merri.

Ned's father John (born Sean Ceallaigh near Tipperary in Ireland) had been a convict (transported for the theft of two pigs) in Van Diemen's Land, but in 1848 he was freed, and travelled to Port Phillip. The Irish community that had established itself on the windy plains at the foot of the Great Divide, less than twenty years after Hume and Hovell had stood on Bald Hill and praised the very same country, drew John inland, through Epping and Donnybrook and eventually up to Wallan East. In 1850, Kelly senior 'was engaged in splitting and fencing, near the Merri Creek, and chancing to be in a hotel…in the vicinity of James Quinn's residence…he for the first time encountered the latter individual… the meeting was the commencement of an intimacy which finally gained Kelly admission to the farmer's home…'[25]

Later that year, John ran away with James' second eldest daughter, Ellen, and the two were married in Melbourne. On the marriage certificate, both gave as their address the 'Merri Creek'. The young couple returned to the maternal household, and built a hut on the banks of the creek, where two baby girls were born. In 1854, the family moved to their own home, a couple of miles to the west in Beveridge. Little Edward was born a year later, either in a house in Beveridge or more probably at the Quinn homestead. Ned grew up in the Beveridge area, on various family farms, with no doubt much toing and froing across the Merri Creek to visit and stay in his grandparents' lively household, until trouble between the large Quinn clan and the local police induced both the Kellys and the Quinns to move farther north. That was in 1864. Late the following year, poor John 'Red' Kelly, a battler if ever there was one, died at the age of forty-six, and the eleven-year-old Ned, now the male head of the Kelly family, signed the death certificate.

From this point, the story of course rises gradually to the level of legend, though I found to my surprise that I actually knew hardly any of its details until I started reading up on it. But in light of its now revealed link with my own country, recovering the full story had become a matter of urgency for me. It was particularly exciting to me that the legendary figure of these parts should turn out to be Ned Kelly, for, although he is a *white* hero, Kelly has also made his way into the Dreaming stories of some of the Aboriginal peoples of Northern Australia.[26] As one who put up a spectacular resistance to white man's law, and defended those whom that law oppressed, Kelly has been embraced, up in the region of the Victoria River, for instance, as an Ancestor, a mythic exemplar of a larger Law of Life which both black and white can avow. And now, as part of the story of our creek, this wonderful hybrid Ancestor, who holds out the possibility of weaving black and white alike into the fabric of the land, was ours!

Really, what *didn't* this obscure little tributary offer the pilgrim?! Temples, divine colossi, medieval gold-spired churches, festivals dedicated to sacred birds, flora parks and sanctuaries and holy waters, and in addition a string of powerful originary sites! What was going on? It was not as if we had chosen this creek as the site of our pilgrimage because we knew it to be particularly significant. We had chosen it simply because it was *there*. It happened to be part of our local environment.

It was not until years after I personally had first bonded with the creek that I started to turn up its unsuspected links with my own family history. My great grandfather had settled on the flood plain in Brunswick in the 1860s, and my grandfather and father had been born in the creek's environs, close to where I currently live. The family had moved to the south bank of the Yarra in the 1920s, and not returned to Merri country until the 1960s.[27] I had not known any of this history when I settled down to raise my own son in Brunswick. But the poetic significance of the creek had only continued to grow since I first became entwined with it.

What kind of magic was this? Would *anywhere* prove equally rich in significance once one had made it the focus of this kind of attentiveness, once one had become wedded to it, as 'country'? And pilgrimage? Is pilgrimage a kind of philosopher's stone that can open up the mythic inexhaustibility of reality to the pilgrim heart, however seemingly banal the route and routine the destination?

After our lunch in the rain at the back of Walnaring, we returned to the banks of the creek. Our little waywender had from here been taken in hand and straightened out, its lazy bends removed and untidy banks smoothed. It parted the paddocks now in a well-behaved, clean-cut fashion, though admittedly under cover of thick gorse, until it reached Wallan East. Little remained here of the picturesque place that Hume and Hovell had named 'Tempe Valley'.[28] 'In this place the Willow Trees has a very beautiful appearance,' Hovell jotted down erratically in his diary; 'it resemble at a distance a Lemon or Orange Tree, The Forest below us extend, and I have no doubt but it will open into a very fine Country – the Creak run to the southward – The Native Flax Grow in abundance, it appear where the ground is good, that it is a common weed, and if cultivated it would be equel to the best that is grown at home, we found that the Bronze Wing Pigeon, the King Parrot & the Native Dog are common.'[29]

Our own approach to Wallan East was uneventful, our day's travel, minus meanders, being quickly accomplished. To what end? We arrived at the tiny cluster with most of the afternoon still ahead of us. But what in the whole world could compare with moseying through the countryside at a creek's pace? What could substitute for the lure of the unexpected, the bracing discipline of pilgrimage?

Certainly not the heart-sinking banality of town life, sampled in a pizza parlour in the main town of Wallan, on the far side of the Hume Highway. This attempt to find solace in coffee and cake dampened our spirits more thoroughly than a whole month of mornings trudging through drizzling paddocks could have done. Listlessly, we wandered back to Wallan East and flopped in the pub, perking up a little when some of the locals volunteered recollections of the creek in flood.

Late in the afternoon, Greg and Kari arrived with our gear and food, and we were finally able to return, gratefully, to our story, making our way up to Bill's paddock. Bill lived at the far end of the little strip of houses that was Wallan East. Wallan East was marooned on an island between the fierce torrent of the highway and the train line: its reason for existence was the railway station. Bill had grown up here, and remembered with immense affection his childhood adventures on the creek – which now lay on the far side of the railway tracks. His grandfather had built the original house and the family had occupied the property – adding dwellings from time to time – for a hundred and twenty years, since first selection. Bill had agreed to let us camp the night in his rail-side paddock.

We found a spot in dripping grass beside a dam full of very happy frogs. The campsite was a scene of extraordinary activity as we unpacked. Goods vans and passenger trains passed incessantly, blasting their whistles, and a level crossing which led to a broken bridge flashed its lights and rang its bells. To the north, just beyond a screen of scrub, the Hume swept past. With an aerodrome also operating on the other side of the highway, we were hemmed in by the roar of heavy vehicles. A rather surreal scenario for pilgrims! The clamour of the frogs, however, made for a nice counterpoint to the trucks, trains and railway bells.

Once night fell, the camp assumed a cinematic aspect, with a blazing gypsy fire in the foreground and the flashing confusion of lighted carriages and black goods vans against the sky behind. Only a bluesy soundtrack was lacking. The rain eased, we dried our wet socks against the flames, and a warm conviviality sprang up in the midst of chaos. When we retired, I slept with my ears tuned exclusively to the frog band on the sound spectrum.

The last morning of our pilgrimage dawned gold and remained so, only gaining in thoroughly biblical radiance as the day wore on. I was on the lookout now for an Indian gentleman I had encountered on a previous visit with Cin. Looking for a campsite, we had driven up to the level crossing and discovered the collapsed bridge. (A sign announced that it had not been in use since 7 September 1989.) Taking a liking to the spot, we settled down for a picnic on the creek bank. Just as we were biting into our feta cheese and corn thins, an Indian personage of breathtaking beauty appeared on the opposite bank, flanked by his two young sons. I rubbed my eyes. What was this? What was Krishna doing here out the back of nowhere, by a broken bridge? But he was perturbed, in a very mortal sort of way. What were we doing on his land, he wanted to know, polite but testy. We explained our mission. Could we cross his property? Too much gorse, he said. Later, after I had sent out a letter to landowners there in the upper reaches, he telephoned. Definitely no, he said. Too many snakes! But still, he added, God bless!

As fate would have it, the full gypsy horror of our camp was now pitched just metres from his boundary, and I imagined him observing us from the upstairs window of the suburban castle that stood in the midst of his little kingdom of gorse. He did appear, of course; as we set off down the railway tracks, bypassing his property as he had requested, we could see him chasing a goat in front of his house, his face turned towards us.

Two other owners had refused us permission to cross their land immediately north of Wallan, so we followed a vehicle track along the railway easement instead, which fortunately ran in tandem with the creek here for as long as it mattered. We paused to rest in a wildflower garden on the railway verge. Yellow bulbine lilies, milkmaids, chocolate lilies, rice flowers – a feast of tiny beauties unfurled out of the raw clay.

Now that we were in foothills, the river reds had finally made way for other gums – swamps and messmates, perhaps, though our botanists were not entirely sure. Before the track disappeared under reeds, Cinnamon alighted on a gold ring in the mud. I was ravished by this extraordinary token, but the gift and its message were for her, not me, so I held my peace.

When the vehicle track gave out, we were forced to walk for a little distance on the railway line. We were heading northwards, but the line was also carrying northbound traffic, rather than southbound, as we had for some reason assumed. It was only because there happened to be another level crossing up ahead, which started ringing and flashing, that we realised that a train was bearing down on us silently – and very rapidly – from behind. We just had time to leap from the tracks when the monster was upon us, with an avalanche of noise. Maya squatted only a couple of metres from the wheels, her arms over her head, her hat blown off! Amidst all my worries about bulls, rogue cows, territorial dogs and the rest, being flattened by a train was one fate I had not anticipated!

After the level crossing, we returned to our riparian haunt, and found ourselves in a new kind of bush idyll – lightly wooded hills with sunny lawns and meadows browsed by sheep. The creek was here the merest rill in the bed of a wide, grassy channel, shaded by swamp gums and messmates. The road, railway line and watercourse criss-crossed fairly frequently here, and the creek was lined, for stretches, with big forest, too thick to enter. Small farms proliferated and, although this more settled landscape lacked the moody appeal of the flatlands, I took comfort from the relative density of bird life that rural subdivision seemed to entrain. We were joined at one point, too, by three tiny yellow dogs, who gambolled and frisked with us through a long dappled paddock, adding that dimension of animal companionship which was the one thing I had really missed on our journey. This, by the way, was our sole canine encounter – three little lap-loving beasties in place of the fierce farm avengers I had dreaded!

Tributaries were now busily joining the creek from different directions, as we were well into the region of the headwaters, but we were resolved to follow what appeared, at each juncture, to be the major channel. This major channel now ceased heading north, and struck out to the east, passing several lily-choked billabongs and tunnelling under the railway line. (Even Maya was not brave enough to broach the hip-high water in this gloomy pipe.)

On the other side of the line, the landscape opened into a wide valley. This was presided over by a large suburban house set magisterially just beneath the wooded ridge top, and adjoined by a workshop enclosure. We spread our lunch on the banks of a deep clay gully in the direct line of view of the house but, as in all previous cases, despite signs of occupation, no one stepped out either to greet or to challenge us. Even when, following the gully a little higher, we came to a capacious dam, and Maya and Cinnamon stripped off and, amidst ear-splitting squeals, plunged into the chilly water, no one appeared.

By this stage the watercourse had almost disappeared, and a couple of step-wise dams led up to a higher paddock, where it petered out altogether. A sturdy gum and a patch of clover with a snake in it seemed alone to mark the Source. A scattering of dumped cars on an adjacent spur set a final seal of disappointment on the scene.

From the very start of our journey, I had had a certain ceremonial intention with respect to the Source, but to carry it out seemed unimaginable on this exposed slope. Although I accepted that this was it, and that in this surreal juxtaposition of the sacred and the crass, the ambiguity of the creek's story was contained, I felt a kind of unthinking agitation to get to the treeline, and after a pause we hurried to the top of the ridge.

At the sight of what lay on the other side, my heart turned cartwheels! There, in thoroughly sylvan seclusion, was a glittering lake! A quick perusal of the map revealed that another tributary, which we had missed down below the scowling house, led up to this, so the lake was ours! The grand nineteenth-century explorer, Sir Richard Burton, could not have been more overjoyed when he discovered Tanganyika, the great waters that fed the Nile, than we were to discover little Pilgrims Lake. We scrambled down to its southernmost shore, and now we *all* stripped, and, in a gilded moment, dived into the Source. For several minutes we swam and circled, our limbs disappearing in the tarnished amber depths, then we returned to the bank and sat, glistening with the Merri waters, like three newborns.

Thus baptised, and flanked by my two dear pilgrim witnesses, I ceremonially asked the creek, its guardian spirits and the ancestors who still dwell within its rocks and clefts, if I could have the honour of bearing its name, to announce our interlocked identities, and to seal my custodial pledge. After I said the words, Cinnamon noticed, out of the corner of her eye, the form of a diver bird, but we could not be sure it was a kingfisher. The sun clouded over, but I sat there, naked and glistening after my christening, my heart pealing with gratitude. One could never have hoped, in one's most optimistic moments, that the Source would turn out to be so generous, so luminous, so magical, hidden as it was – behind a facade of profanity – on the other side of reality. How could the story be so true? If the humble Merri could offer this to its pilgrims, what could the wider world offer its people, once they approached it as devotees? The vision-filled lakes of old Tibet, perhaps? The oracular landscapes of the classical world? Who could know…

And who else knew this secret of the Source, the significance of this little sister to Lake Tanganyika, nestled in the foothills of the Great Divide? The people in the house over the ridge? Not likely, judging from appearances. The staff of the Creek Committee? If so, why had they not mentioned it? Those who had drawn up the ordinance map would of course have been aware that the Merri was fed by a sizeable body of water, but what would this have mattered to them? From a cartographic point of view, it might just as well have risen amongst the dumped cars. The cartographers had not grabbed the creek by its lapels and, staring into its eyes, asked, Who are you? They had not trudged for seven days on rotting feet to find an answer to the question. The Merri, the Darebin, White Elephant Gully, Bruces Creek, the Dry, all these would have been the same to them, parts of an undifferentiated externality, laid out, waiting to be measured, fitted, transcribed, devoid of inner meanings. Suspecting no secrets, failing to address, speaking only to each other over the heads of the landforms they were mapping, the map-makers would of course have received no revelation. Only those who ask, Who are you? and faithfully trudge the trail to find out, do so. Only they get to see the other side of cartography.

So does it really matter, in the end, if wildernesses and special sites are profaned by the irreverent eyes of the casual, the curious, the touristic? Are not quests for the numinous and unknown always, in essence, quests for revelation, and is not revelation necessarily vouchsafed only to those who come to the world as seekers and supplicants, speakers of a subterranean poetics? A site which remains impenetrably veiled in banality to the hundreds of passers-by to which it might be daily exposed may nevertheless choose to unveil itself – in a uniquely revelatory fashion, via dreamlike conjunctions and sequences of circumstances – to the gaze of the initiate. The sacred is surely always in this sense hidden from the eyes of the profane, and who but the initiate can say what is sacred and what isn't? By the same token, even the most exotic sights and sounds encountered by colonial explorers might have failed to count as true discoveries, true inroads into the unknown, if they were perceived simply as part of the same great empirical externality which had already been so fully charted.

Our own expedition, in contrast to these colonial travels, was a journey, not *over*land or *around* the world, but *into* the world, *into* the land, a walking *through* the appearances into a terrain of speaking that lies beneath those appearances, that utters them.

Isn't all in-land journeying, then, a journey to a primal source? And couldn't we rest assured that, while other seekers might make their own way in-land, the secret of the Source in this instance, in the shape of this little sister of Lake Tanganyika, as it had been given to us, was forever ours?

The lake itself naturally had its tributaries, and one of these rose in the glade behind our swimming place. At first it consisted of just a pool here and there in a wide, marshy watercourse, but as we climbed into the thickly wooded gully, a rivulet took shape, varying from weedy runnel to springy flat to deep tunnel of erosion in the clay. Here, the sunlight was strobed by straight trunks, making zebra patterns on the slopes. The ground was whiskered with moss and maidenhair, and set with tiny venus fly traps. At one point, the valley filled with mysterious wooden pallets nailed vertically to trees. After Cinnamon found a couple of green paint balls in the mud, we realised that this area must be used for military training or war games. Suddenly, the slanted shadows seemed astir with ghosts and echoes, phantom heads popping out from behind boards, incongruous shouts breaking the intense stillness. We took this in our stride, as the creek did, and passed on. It was no more bizarre than any other episode in the creek's story.

At this point, however, a lone horse loomed up from the depths of the forest. Actually, there were two horses, but one of them was so old and wasted, standing absolutely motionless and matt amongst the shadows, it could have been a dead horse, propped up. But the other was all too alive and immediately started to shove and harass and push at us. We had already been through this horse business once before, at Walkers Farm, and my heart began to pump, but fortunately a fence appeared out of nowhere, and, contortionists that we had now become, we were through in a flash.

It was time now to fill our flasks. We lay, cheek to the leechy mud, and reached down into a cavity in which water was audibly running. We waited for the trickle to fill our cups, then drew them up. Reaching out to our land in this last tender sacrament, we drank. The water had no taste, but was icy, and clear as Perrier. It was hard to adjust to this fact. The unpotability of the Merri had always been part of our conception of it. This intimate act of drinking had been ruled out – undreamed of – from the very start. How this taboo must permeate our thought! What is it to walk in a world in which the waters are poison and food nonexistent? How can one belong, in any but a sentimental way, to such a world? But now we were breaking through the bindings of this taboo. We reached our flasks down and collected a little of the miraculous draught to carry home.

Soon the orchids began to appear. Cinnamon and Maya found them, of course, and I stooped, speechless, to admire these holy grails of the wildflower realm. Several different kinds grew up the sides of the gully. There were creamy bridal ones, and others that were port-red and green-haired, squatting open-legged, like little chthonic goddesses, close to the ground. Further up, we were greeted by sky-blue, wide-eyed faces on high stems. What could one say? Could Eden itself, where the primal rivers rose, boast anything more holy than these tiny illuminations from the manuscripts of creation? I had never imagined, never dreamed of receiving, such an access of graces, and I trailed behind my two companions, brainstruck, my head ringing like a Tibetan singing bowl.

The sun was setting. It was time to rendezvous, up at the mountain crossroads, with Greg. We had found the Source. We had swum in it, walked in it, drunk of it. Our hair was still wet with it. We had entered it, and it had stepped forward and received us, kissing us each on both cheeks. For myself, I knew that I belonged to it. There was no turning back. The Tao rises, in infinite beauty, shrouded from view but revealed, in the midst of desecration, to the pilgrim. There are no words for this. There are no maps for this. Everyone invents their own journey to the Source, to find their own revelation and rebirth. The Source, as all the guidebooks agree, is everywhere. To find it one has only to ask the right question. But this is a question that has never yet been asked, because it is uniquely one's own. The Tao flows, patiently wending its way around all-that-is-named, down to the great sea of namelessness to which every last thing returns.

I would like to offer heartfelt thanks, on my own and Maya's and Cinnamon's behalf, to all who helped us to reach the Source of the Merri. Our support team: Len O'Neill, Noel Blencowe, Viv Rodknight, Cathie Nixon, Thais Sansom and family, Gerard Farmer, Saer Lachey, Greg Milne and Kari. Our hosts: the Coburg Scouts Club, the Campbellfield Community Centre, Margaret Walker, Jerry and family of Donnybrook, Kate and Stephan Adey, Bill of Wallan East. The Launch at CERES: Tony Faithful of MCMC, Ray Radford of the Friends of the Merri and Alan Reid of the Gould League.

We acknowledge our incalculable debt to those who belonged to the creek before us, and belong to it still, and who are our teachers and elders in creek matters as in so many other respects, the Wurundjeri people, both ancestral and of the present day.

Notes

1. I am of course not alone in harbouring such deep childhood memories of a creek. The childhood significance of creeks and rivers, even for city dwellers, has been documented by many writers. (See for instance Thomas Campanella, 'The Lost Creek' in Terra Nova 1, 4, 1996, 113–119, and Robert Pyle, The Thunder Tree: Lessons from an Urban Wildland, Houghton Mifflin, Boston, 1993.) According to Peter Steinhart, most people have a creek in their deepest memories. 'Nothing historic ever happens in these recollected creeks. But their persistence in memory suggests that creeks are bigger than they seem, more a part of our hearts and minds than lofty mountains or mighty rivers. Creektime is measured in strange lives, in sand-flecked caddisworms under the rocks, sudden gossamer clouds of mayflies in the afternoonÖ Mysteries float in creeks' riffles, crawl over their pebbled bottoms and slink under the roots of trees.' (Peter Steinhart, 'The Meaning of Creeks', *Audubon*, May, 1989, pp. 22–23.)

2. As I was revising this essay for publication, I came across a recently published book entitled *Waterlog* by Roger Deakin, Vintage, London, 2000. Deakin offers an account of just such an adventure into the heart of the given: having undertaken to swim his way across Britain, he searched out rivers, swimming pools and water holes of every kind, embarking on a watery quest wilder at heart than any eco-tour into South American cloudforests or the Siberian taiga.

3. Johann Wolfgang Von Goethe, *Faust* Part II, translated by Phillip Wayne, Penguin, London, 1959. p. 258.

4. John Anderson, *The forest set out like the night*, Black Pepper, Melbourne, 1997. p. 13.

5. 'Merri' is the Wurundjeri word for 'rocky'.

6. His full name was the Venerable Thich Tinh Dao.

7. Michael Lafargue, *The Tao of the Tao Te Ching: a Translation and Commentary*, State University of New York Press, Albany, 1992. p. 102.

9. Ibid. 16.

10. Richard Wilhelm (trans), *I Ching*, Arkana, London, 1967. p. 115.

11. Lafargue, op. cit. p. 138.

12. See Helen Penrose (ed.), *Brunswick: One History, Many Voices*, Victoria Press, Melbourne 1994, ch. 1.

13. Anderson, op. cit. p. 14.

14. Alan E.J. Andrews (ed.), *Hume and Hovell 1824*, Blubber Head Press, Hobart, 1981. p. 203.

15. See Kate Rigby, 'The Politics of Pilgrimage', *PAN (Philosophy Activism Nature)*, no. 1, 2000.

16. Rigby makes this point. 'There is no necessary link between pilgrimage to...culturally co-constituted sacred places and an enhancement of ecological consciousness, let alone eco-political activism. However, where pilgrimage is reconfigured as a journey in which we honour, not only particular cultural mediations of the sacred, but also and especially the underlying order of ecocosmic creativity within which such mediations are enfolded and by which they are sustained, then it might indeed contribute to the rehallowing of the whole Earth, and to a renewed commitment to its continued flourishing.' Ibid. p. 30.

17. We were 'walking the world into being', to use the wonderful title expression of an episode of the ABC Radio National program *The Listening Room*, broadcast in 1991. (No other details available.) Satish Kumar also emphasises the importance of such step-by-step contact in his spiritual autobiography, *No Destination*, Green Books, Devon, 1992. 'Pilgrimage is best,' he says, 'when you put your body on the pilgrim's path. The Tibetans go on pilgrimage by prostrating themselves every inch of the way. They start by standing, their hands held in prayer, then they kneel and, bowing down, lie facing the earth and touching her with their forehead in humility. Then they make a mark on the ground with their nose, stand up and walk up to the spot marked by the nose (to ensure that they do not miss a fraction of the path). Then they stand and repeat the process, and this they do all the way to the temple, which

might be one hundred miles away.' (p. 176) I was moved by Kumar's account of a pilgrimage he had taken around the sacred sites of Great Britain. Kumar was born in India and spent his early life as a wandering Jain monk. It is apparently an Indian tradition that in one's fiftieth year one makes a pilgrimage to the holiest Hindu and Buddhist places of India. (I was gratified when, later, I remembered this Indian tradition, for the Merri walk had fallen, not at all by design, just after my fiftieth birthday.) By the time Kumar himself reached fifty however, he had long been resident in Britain, and was settled in Devon with a British family. Undeterred, he made his pilgrimage to the holy – Christian and pre-Christian – places of Britain instead. He describes the intention behind his four-and-a-half month walk as follows: 'In India, before you enter a temple you go around it. There is a precinct for that purpose, and by going once, twice, three times round you prepare and centre yourself. You leave your negative thoughts behind. When your body, mind and heart are ready, then you may enter the temple. Similarly, I am making a journey around the temple of Britain, so that I may enter into its mysteries. This pilgrimage is a pilgrimage to Britain, to its rivers, hills, moors, dales, fields, to all its natural beauty.' (p. 180) In the early 1960s, Kumar had embarked on an even more impressive journey, a 'peace march' from India to the four nuclear capitals of the world, Moscow, Paris, London and Washington. He and his companion travelled only by foot, at least where the route lay overland, and, most astonishingly, they carried no money. Reflecting on the eighteen month walk, Kumar wrote, '[i]n wandering I felt a sense of union with the whole sky, the infinite earth and sea. I felt myself a part of the cosmic existence. It was as if by walking I was making love to the earth itself. Wandering was my path, my true self, my true being. It released my soul-force, it brought me in relation to everything else.' (p. 110) As a result of his walking around the world, it seems that the planet itself became 'country' for Kumar.

19. Discovering new stories or revisioning old ones is essential if the power of stories to unify the land is to be maintained. Roger Deakin's swimming journey through Britain is a beautiful contemporary example of revisioning country, perceiving it through a new poetic frame and in that sense rediscovering it, repossessing it.

19. For an account of the way in which Aboriginal songlines or 'strings' function as lines of both discrimination and connection simultaneously, see Deborah Bird Rose, *Dingo Makes Us Human*, Cambridge University Press, 1992.

20. When the role of waterways as creators of country is appreciated, the ancient Greek perception of rivers as gods becomes more comprehensible: as an agent of creation – poetic as well as physical – the river has an undeniably mythical status. (See Harry Brewster, *The River Gods of Greece : Myths and Mountain Waters in the Hellenic World*, I.B. Tauris, New York, 1997.) Why not honour it then in a mythic idiom – by deifying or otherwise personifying it? How might the Merri respond if we addressed her as a goddess?

21. Kate Rigby brings out the complexity of 'pilgrimage' in colonised lands in 'The Politics of Pilgrimage', op. cit.

22. See Robyn Davidson, *Desert Places*, Viking, New York, 1996, for an account of the history and culture of one group of such peoples.

23. A significant step in the politics of repossession and reinhabitation would be to campaign for public access to all the waterways in our state and ultimately across our continent. Waterways are, as I have remarked, the natural pathways through the land, and if people are to 'walk their world into being', then a primary desideratum is that they should have access to the waterways. Such access has apparently been legislatively assured in New Zealand: a colonial law affording a 22-yard strip of public access along the banks of every river in the country has recently been re-enacted. (The law is known as 'the Queen's chain' because it was originally requested by Queen Victoria.) See Roger Deakin, op. cit. p. 34.

24. In August 2000, the Victorian National Parks Association and the Friends of the Merri put forward a proposal to encompass the entire length of the Merri creek, up to Wallan, in a linear park. The park would be 2,500 hectares in area and would link all the significant biological and geological sites along the creek.

25. G. Wilson Hall, *The Kelly Gang: the Outlaws of the Wombat Ranges*, G. Wilson Hall, Mansfield, 1879. Quoted in Keith MacMenomy, *Ned*

Kelly: the Authentic Illustrated Story, Currey O'Neill Ross, Melbourne, 1984. p, 11.

26. Deborah Bird Rose, op. cit.

27. After I published a little account of our journey in a local newspaper, I received a phone call from a man who wanted to know what had first attracted me to the Merri Creek. It turned out that he was a genealogy buff, and amidst his researches into his family tree he had come across a forefather of his, one George Mathews, who had been swept off the bridge that crosses the creek at Heidelberg Road during a flood in 1854. George's body had never been recovered, so his bones, or at least his ghost, presumably still lay in the bed of the creek. Was I related to George? I don't know. The Mathews branch of my family were arriving in the area at around that time, so George may well have been an uncle or brother of my great grandfather.

28. This was no doubt named after the sacred Tempe valley in ancient Greece. See Harry Brewster, op. cit.

29. Andrews. op cit. p. 201.

www.ingramcontent.com/pod-product-compliance
Lightning Source LLC
Chambersburg PA
CBHW030916080526
44589CB00010B/340